MW01714906

Desert

Daniel Moreton

Scholastic Inc.
New York • Toronto • London • Auckland • Sydney

Acknowledgments

Science Consultants: Patrick R. Thomas, Ph.D., Bronx Zoo/Wildlife Conservation Park; Glenn Phillips, The New York Botanical Garden

Literacy Specialist: Ada Cordova, District 2, New York City

Design: MKR Design, Inc.

Photo Research: Barbara Scott

Endnotes: Samantha Berger

Endnote Illustrations: Craig Spearing

Photographs: Cover & p. 1: Stefan Lundgren/The Wildlife Collection; Cover type: Peter Arnold; Interior border: Stefan Lundgren/The Wildlife Collection; p. 2: Bruce M. Herman/Photo Researchers; p. 3: Thomas Dressler/DRK Photo; p. 4 & 5: Tom Bean/DRK Photo; p. 6: Renee Lynn/ Photo Researchers; p. 7: James P. Rowan/Tony Stone Images; p. 8 & 9: Larry Ulrich/DRK Photo; p. 10: Charles Melton/The Wildlife Collection; p. 11: Michael Francis/The Wildlife Collection; p. 12: Jim Steinberg/Photo Researchers.

No part of this publication may be reproduced in whole or in part, or stored in a retrieval system, or transmitted in any form or by any means, electronic, mechanical, photocopying, recording, or otherwise, without written permission of the publisher. For information regarding permission, write to Scholastic Inc., 555 Broadway, New York, NY 10012.

Library of Congress Cataloging-in-Publication Data
Moreton, Daniel
Desert / Daniel Moreton.
p. cm. -- (Science emergent readers)
Includes index.
Summary: Photographs and simple text explore the environment of the desert, including its sand, rocks, cacti, and canyons.
ISBN 0-590-63869-6 (pbk.: alk. paper)
1. Desert--Juvenile literature. [1. Deserts.]
I. Title. II. Series.
GB611.M665 1998
551.41'5--dc21 98-23225
CIP AC

Copyright © 1998 by Scholastic Inc.
Illustrations copyright © 1998 by Scholastic Inc.
All rights reserved. Published by Scholastic Inc.

Printed in the U.S.A.

7 8 9 10 08 03 02 01

In the desert...

there is sand

and more sand.

There are rocks

and more rocks.

There are cacti

and more cacti.

There are flowers

and more flowers.

There are mountains

and more mountains.

There is water—but only a little bit.

Desert

One of the harshest places in the world to live in is the desert. The temperatures are severe, and there is very little rainfall. A place is called a desert when there is less than 10 inches of rainfall in one year. That's not even enough water to fill up a bathtub! What does the desert look like, and what might you find there? Here are some of the things you would most likely discover on a trip to the desert.

In Monument Valley in Arizona (page 1), the long stretches of sand have been blown repeatedly by wind, making these ridge dunes, which are like waves or ripples in the sand. This is one of the most well-known settings in the world, having appeared in many western films. Like most deserts, Monument Valley has plenty of sand.

The sand dunes in Death Valley, California (page 2), were formed by the wind as well. Death Valley is the hottest and driest part of North America. The temperatures there are some of the hottest on earth! Here, sand dunes stretch as far as the eye can see.

Sand dunes in the Namib Desert in southwestern Africa (page 3) are some of the oldest in the world. Below the dunes that are there today lie ancient dunes dating back some 40 million years! Both the old and the new dunes were formed by strong winds carrying sand from the coast nearby. Some of the Namib dunes are 800 feet high!

Many years of erosion and exposure to harsh climates have caused the rock formations in the Painted Desert (pages 4–5) to become a multi-colored wonder. It's common to see stately, ruddy, rusty red rock

formations like these in many deserts in the American West. Iron that has been oxidizing in the rocks for years and years is responsible for the brilliant colors that now exist.

People often think that plants and flowers cannot grow in deserts, but that is not the case. The golden barrel cactus (page 6) is one of many cacti that can be found in the desert. They tend to grow in the canyons where the environment is warmer and more moist.

The prickly pear cactus (page 7) is one of the most common cacti and covers many square feet of desert. It stands 3–6 inches in height and is a flat, almost oval cactus. In the late spring, bright yellow and magenta flowers may bloom.

People journeying to the desert often try to plan their visits sometime during the spring or early summer. That is when the cacti and flowers are in full bloom and the desert is filled with splendid color! Lambert's locoweed and yellow cryptanth (pages 8–9) are just one example of the blooming flowers one might find on a trip to the desert.

Many deserts contain mountainous structures, like the Anza-Borrego Desert in California (page 10) and the Badlands in South Dakota (page 11). For hundreds of miles, incredible hills and mountains rise up from the earth, like an ancient lost city, to form these massive peaks. For half a million years, water has been eroding these cliffs. Sometimes water will carve away over an inch of cliff each year! Often the erosion uncovers ancient fossils of creatures who lived there thousands of years ago.

One thing common to all deserts is the scarcity of water (page 12). When it does occasionally rain, water saturates the ground and filters down through the rocks. Certain rocks hold water and are called aquifers. The water held in an aquifer may be thousands of years old!